YOUR KNOWLEDGE HAS VALUE

- We will publish your bachelor's and
  master's thesis, essays and papers

- Your own eBook and book -
  sold worldwide in all relevant shops

- Earn money with each sale

Upload your text at www.GRIN.com
and publish for free

Amalia Aventurin

Specific Surface Area, Langmuir, BET

GRIN Verlag

Bibliografische Information der Deutschen Nationalbibliothek:

Die Deutsche Bibliothek verzeichnet diese Publikation in der Deutschen National-
bibliografie; detaillierte bibliografische Daten sind im Internet über http://dnb.d-
nb.de/ abrufbar.

Imprint:

Copyright © 2013 GRIN Verlag GmbH
Druck und Bindung: Books on Demand GmbH, Norderstedt Germany
ISBN: 978-3-656-64477-4

This book at GRIN:

I. Specific Surface Area, Langmuir, BET

Abstract:

The specific surface area it is a physically and chemically important parameter of porous materials. It is most commonly determined by gas sorption isotherms. Two of these – The Langmuir- and BET-isotherms – are evaluated in this exercise.

I.1. Introduction

The theme of this report is the evaluation of gas adsorption isotherms using the Langmuir, BET methods to assess the specific surface area. Gas adsorption experiments on porous materials yield different isotherm curves. These curves have been classified into six types by the International Union of Pure and Applied Chemistry (IUPAC) (Figure 1).

Adsorption is in general an enrichment of gas or fluid at a surface/solid, in more detail, at the border of two phases. Because of the complexity and difficulty of adsorption and desorption, process of leaving atoms or molecules from the surface of a solid, the data get analyzed by the following Isothermal models. These models assume dynamic equilibrium at constant temperature between adsorption and desorption, and are described in more detail in the following topic "1.2 Theoretical Background".

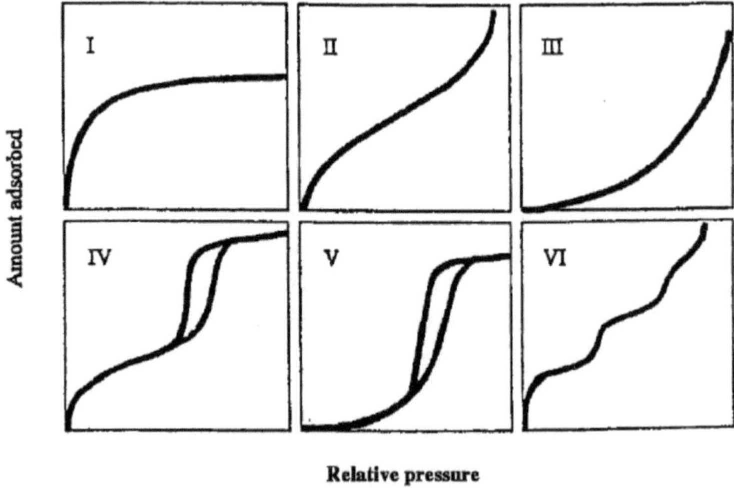

Fig 1: The six different types of adsorption-isotherms according to the IUPAC classification.

The type I isotherm in Figure 1 is the Langmuir-isotherm. This isotherm type assumes monomolecular sorption and approaches limiting value at high pressures as shown in the sketch.

Type II is the BET-isotherm. This situation appears after the development of the monomolecular cover furthermore a cover of fluids, so that there is at the end a multi-layer system.

Type III is a curve which appears if there is only a small force between the first layer and the solid. Because of this reason, the development of the multi-layer system is finished first before the monomolecular layer is completed.

Type IV and V are special curves which occurs only at capillary condensation. Type V has the same phenomena like type III, at both types are small forces at the beginning. The types III and V are very rare because of the small interaction force at the adsorption [4].

Type VI is a very special process; it shows a stepwise multi-layer adsorption. This case is the rarest and because of this neglected in the report.

The most important types are I and II and on these two types we take a closer look. At first we go into detail to the Langmuir-isotherm and after that to the BET-isotherm and the specific surface area.

I.2. Theoretical background

I.2.1. Langmuir isotherm

The Langmuir isotherm (see Fig. 1 type I) is a mathematical way to describe monomolecular adsorption. This model shows the variation of adsorption with pressure. It gets proposed by Irving Langmuir in 1916 [5]. This adsorption isotherm is based on the following assumptions:

- There is a dynamic equilibrium between adsorption (adsorbed gaseous molecules) and desorption (the free gaseous molecules).
- The adsorptive built on the solid a monomolecular layer (Adsorption is monolayer or unilayer [5]).
- The surface has the same energy level, which can bond one gas molecule. This means that adsorption is localized and the adsorbed molecules can't move on the solid (All the vacant sites are of equal size and shape on the surface of adsorbent [5]).
- No interaction between the adsorbed molecules (Each site can hold of one gas molecule and a constant amount of heat energy is released during this process [5]).
- A fixed number of vacant or adsorption sites are available on the surface of solid [5].

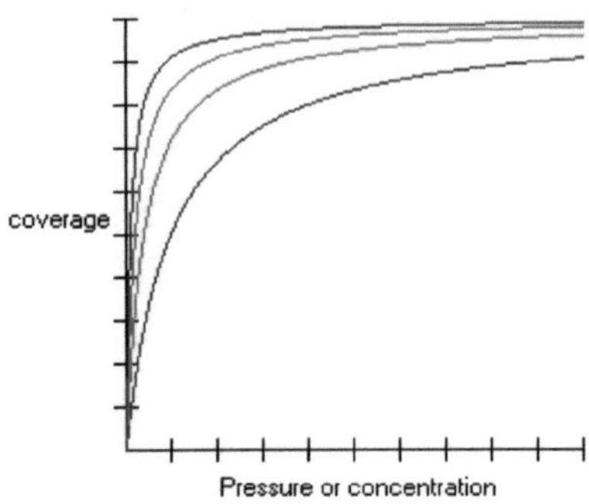

Fig. 2: Langmuir isotherm with different Langmuir constants and identical maximum sorption capacities (coverage) [6].

The Langmuir formula describes the relationship between the fraction of occupied sorption sites (Θ) at a given pressure (P) [5] at fixed temperature. Different curves which show a constant increasing value. [6]

$$\Theta = \frac{b \cdot P}{(1 + b \cdot P)} \tag{1}$$

The Langmuir equation is derived considering dynamic equilibrium between the free gas and the adsorbed gas molecules. Adsorption and desorption rates are in equilibrium as expressed by the formulae below:

Adsorption rate: and Desorption rate:

$$r_a = \frac{\partial \Theta}{\partial t} = k_a \cdot P \cdot (1 - \Theta) \tag{2} \qquad\qquad r_d = -\frac{\partial \Theta}{\partial t} = k_d \cdot \Theta \tag{3}$$

$$r_a = r_d$$

In these formula P is the gas pressure, k_a and k_d are rate constants for adsorption and desorption and Θ is the fraction of occupied adsorption sites. Where Θ is defined as $n_{occupied}$ / g (sorbent) divided by n_{total} / g (sorbent). The equilibrium is then:

$$k_a \cdot P \cdot (1 - \Theta) = k_d \cdot \Theta \tag{4}$$

$$\Theta = \frac{\left(\frac{k_a}{k_d}\right) \cdot P}{\left(1 + \frac{k_a}{k_d} \cdot P\right)} \tag{5}$$

Rearrangement yields the following expression for Θ:

$$\left(b \equiv \frac{k_a}{k_d}\right) \tag{6}$$

By defining b as the ratio of k_a and k_d

$$\Theta = \frac{b \cdot P}{(1 + b \cdot P)} \tag{7}$$

To obtain:

$$\Theta = \frac{m_{ads}}{m_\infty} \tag{8}$$

The fractional occupancy can be expressed as:

$$\Theta = \frac{m_{ads}}{m_\infty} = \frac{P}{K_L + P}$$ where m_{ads} is defined as $m_{adsorbed}$ per (unit mass of sorbent) and $m\infty$ is defined as $m_{complete\ saturation}$ (per unit mass of sorbent). Alternatively the

Langmuir equation can be written as

$$K_L \equiv \frac{k_d}{k_a} \tag{9}$$

K_L is commonly referred to as "Langmuir pressure".

In the experimental method this formula is used in a rearranged form:

$$m_{ads} = m_\infty \cdot \frac{P}{K_L + P} \tag{10}$$

The calculated solutions are listed in the experimental part.

I.2.2. Specific Surface Area and BET

The specific surface area (in short: SSA) denotes the total internal surface area [m²] of a porous medium per unit mass [g]. This parameter is of "importance for adsorption, heterogeneous catalysis and reactions on surfaces" [2]. "The SSA can be measured by adsorption using the BET-Isotherm" [2].

The BET-theory was proposed by Stephen Brunauer, Hugh Emmett and Edward Teller, in 1938 and is named after the initials of their surnames.

This theory explains the physical multilayer adsorption of gas molecules on a solid surface. It is an expansion of the Langmuir-theory, which only considers monolayer absorption, and provides a mathematical representation of "type II isotherm" [1]. The authors expand the Langmuir-theory with the following three hypotheses: First "gas molecules physically adsorb on a solid in layers infinitely" [1], second "there is no interaction between each adsorption layer" [1] and finally "the Langmuir theory can be applied to each layer" [1]. With the resulting equation the specific surface area can be determined from the obtained from experimental sorption measurements at low pressures and temperatures (below the saturation pressure of the sorptive):

$$\frac{P}{m(P^{*}-P)} = \frac{1}{m_{mono}\cdot c} + \frac{c-1}{m_{mono}\cdot c}\cdot\frac{P}{P^{*}} \tag{11}$$

Here m is the mass absorbed [cm³/g], m_{momo} is the mass absorbed at complete monolayer occupation, P the pressure, P^{*} the vapor pressure of sorbate at experimental temperature (often written as P^{0}) and c the BET constant, expressed by $c = \exp(\frac{E_1-E_L}{RT})$, where E_1- is the heat of adsorption for the first layer and E_L for the second and higher layers (equal to heat of condensation).

Figure 3 shows a scheme illustrating the multilayer concept of the BET approach.

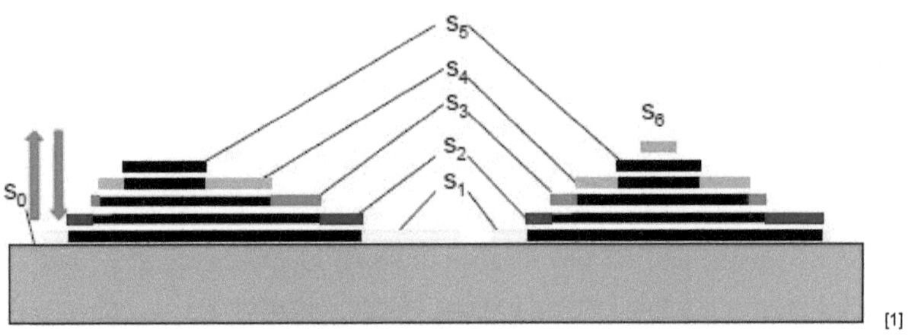

[1]

Fig. 3: Illustration of the BET sorbent layers S_0 is the total surface area- S_1, S_2, S_3, S_4, S_5 represent the surface area covered by 1, 2, 3, etc. sorbent layers. Adsorption $[a_1pS_0]$ is equal to desorption $[b_1S_1\exp(-E_1/RT)]$. "a_1, b_1 and E_1 are independent of the number of adsorbed molecules already present in the first layer". The total surface area (S_0) is constant. The rate of condensation on the bare surface S_0 is equal to the rate of evaporation from the first layer" [1].

An example of a BET isotherm by Brunauer et al. (1938) is plotted in Figure 4. For the calculation of the specific surface area method can just a small part be used, namely the part where the line stays straight. This is not shown in figure 4.

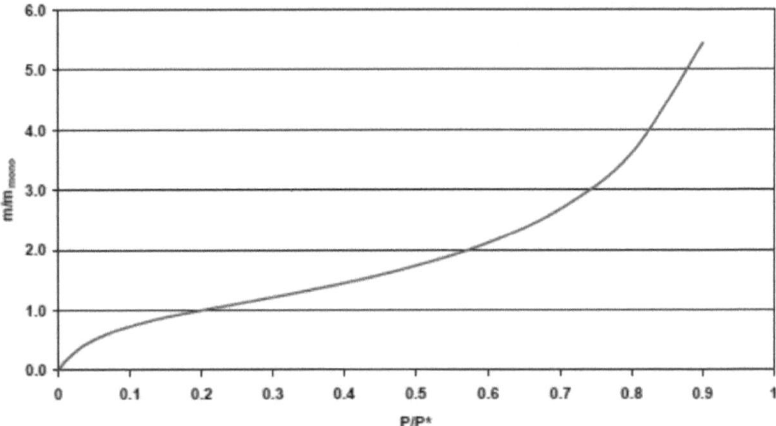

Fig. 4: Ideal BET Isotherm by Brunauer et al. (1938) according to experimental results. On the x-axis is $\theta=P/P^*$ the relative pressure and on the y-axis it is $1/Q[(P^*/P)-1]$, where Q represents the absorbed volume [3].

I.3. Experimental methods and Results

I.3.1. Langmuir

The Langmuir isotherm evaluation is performed with the data provide in an Excel spreadsheet using the following formula:

$$m_{ads} = m_\infty \cdot \frac{P}{P_L + P} \tag{12}$$

This equation can be transformed into

$$\frac{1}{m_{ads}} = \frac{P_L}{m_\infty} \cdot \frac{1}{P} + \frac{1}{m_\infty} \tag{13}$$

providing a linear relationship between reciprocal adsorbed mass and reciprocal pressure. The Langmuir pressure (P_L-) and the maximum sorption capacity (m unendlich) can be calculated from the slope and the intercept of the regression line. Next to this calculation it has also to be calculated the Langmuir sorption along with the error squares and particularly a diagram of the Langmuir-Isotherm. To calculate P_L and m_{inf} the values have first to be optimized, which is done by using the "Sum of squares". This is done by using the sum of the error squares multiplied with 1000.

$$\text{sum of squares} = \text{error squares} \cdot 1000$$

This "sum of squares" value has to be a minimum value (Tab. 1).

Summe of squares	0.14642101	minimum	
P_L	[MPa]		3.18612514
m_{inf}	[mmol/g]		0.464863626

Tab. 1: Excel results of the solver function.

To determine P_L and m_{inf} first an estimate value has to be guessed on the basis of the plotted Langmuir-Isotherm values (see figure 5). This values are solved with the solver-function of excel. The resulting Langmuir-curve out of these values approaches to the fixed guessed values.

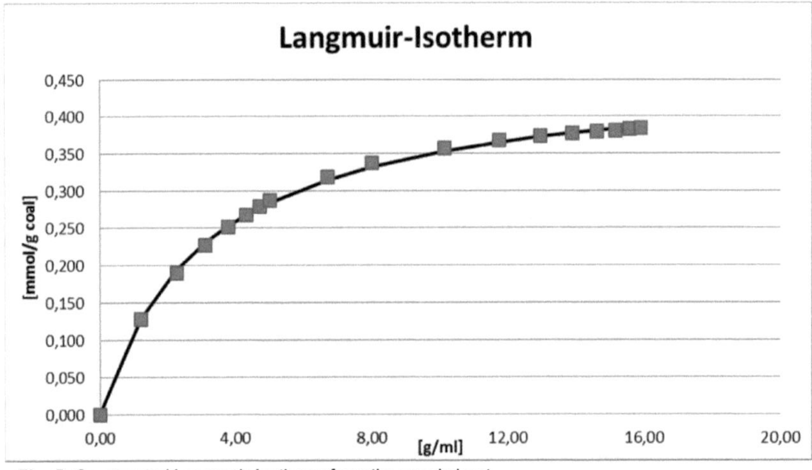

Fig. 5: Constructed Langmuir-Isotherm from the excel sheet

To find a verification, if the guessed values are right, the linear equation is used:

$$y = m \cdot x + b \tag{14}$$

The linear equation is now applied to the following equation:

$$\frac{1}{m_{ads}} = \frac{P_L}{m_\infty} \cdot \frac{1}{P} + \frac{1}{m_\infty} \tag{15}$$

To apply the equation above (15), first $1/m_{ads}$ and $1/P$ has to be determined. The results for this are plotted and show a linear. With this linear, shown in figure 6, you get the intercept and slope. The linear equation of this regression-line is:

$$y = 6.9934x + 2.1334$$

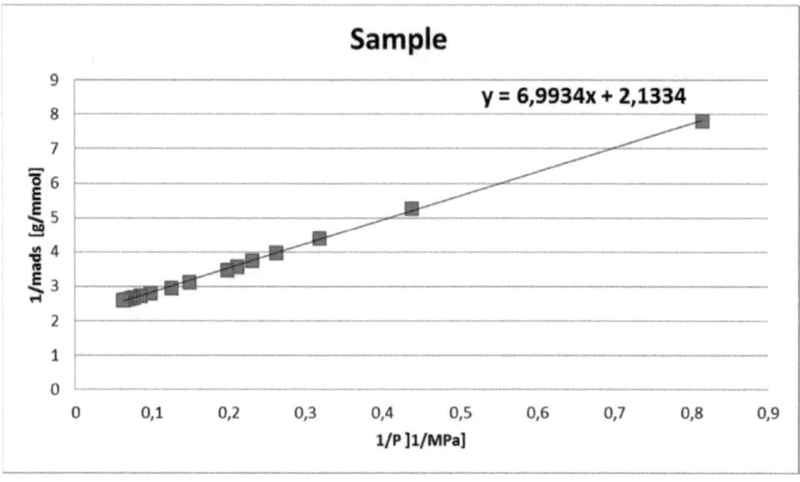

Fig. 6: Regression line to compare the solver values with another calculation.

The intercept of the Langmuir equation is $1/m_\infty$ = 0.46873 mmol/g. Out of the intercept the m_{inf} value can be estimated. The slope represents the value for $\frac{P_L}{m_\infty} \cdot \frac{1}{P}$, where $\frac{1}{P}$ has to be replaced by m_{inf}. P_L can now be replaced by the value of the slope (P_L= 3.27805 MPa):

$$y = m \cdot x + b \tag{16}$$

$$\frac{1}{m_{ads}} = \frac{P_L}{m_\infty} \cdot \frac{1}{P} + \frac{1}{m_\infty} \tag{17}$$

where m_{inf} get calculated:

$$\frac{1}{m_\infty} \text{ is equal to } \frac{1}{m_{inf}} = \frac{1}{2.1334} = 0.46873 \text{ mmol/g}$$

for P_L the calculation is as followed:

$$\frac{P_L}{m_\infty} \cdot \frac{1}{P} \quad \text{where} \quad \frac{P_L}{m_\infty} = 6.9934 \text{ MPa/mmol/g} \quad \text{and} \quad \frac{1}{P} = m_{inf} \text{ (look above)}$$

so one get:

$$\frac{P_L}{m_\infty} \cdot m_{inf} = 6.9934 \text{ MPa/mmol/g} \cdot 0.46873 \text{ mmol/g} = 3.27805 \text{ MPa}$$

The first two resulting values are equivalent to the second two values and therefore the guessed values for the Langmuir-Isotherm are correct. Hence all four measured values match with the given and ideal Langmuir-Isotherm. The second two values also are a check, if the solver-function works right.

The Lagmuir-Isotherm, plotted with the calculated values has the same curve-trend, as Type I, and the values match to the given literature values. This is again an evidence for the right calculation.

For the linear equation the "normal" Langmuir equation $m_{ads} = m_\infty \cdot \frac{P}{P_L+P}$ is transformed into the reciprocal form $\frac{1}{m_{ads}} = \frac{P_L}{m_\infty} \cdot \frac{1}{P} + \frac{1}{m_\infty}$ to calculate the P_L and m_{inf} value with the aid of the linear equation.

For this spreadsheet (Tab. 2) the calculated P_L worth is 3.18612514 MPa (from the Solver) and 3.27805 MPa (from the linear equation). The calculated m_{inf} worth is 0.464863626mmol/g (from the Solver) and 0.46873 mmol/g (from the linear equation). Generally the Langmuir-Isotherm is the simplest sorption model with physical basics.

Pressure corr. [MPa]	Excess Sorption [mmol/g coal] moist	calculated excess sorption (Langmuir sorption)	error squares	1/P	1/mads
0.00	0.000	0	0	-	-
1.23	0.128	0.129339251	1.45174E-06	0.81419964	7.80430748
2.28	0.190	0.19403513	1.7846E-05	0.43807771	5.26840767
3.13	0.227	0.23050371	1.20058E-05	0.31911159	4.40453405
3.80	0.251	0.252810201	3.18268E-06	0.26326181	3.98364803
4.31	0.267	0.267382463	2.70295E-09	0.2318088	3.739234
4.71	0.279	0.277355369	3.0832E-06	0.21218808	3.58280046
5.02	0.287	0.284308818	9.26303E-06	0.1993223	3.48004821
6.71	0.319	0.315208189	1.31326E-05	0.14901575	3.13644729
7.98	0.338	0.33220589	2.91742E-05	0.12533213	2.9620221
10.12	0.357	0.353527855	1.48374E-05	0.09884353	2.79814368
11.73	0.368	0.365551504	5.41969E-06	0.08526894	2.71828119
12.96	0.374	0.373114744	1.46329E-06	0.07717836	2.67147955
13.90	0.377	0.378176932	6.13897E-07	0.071944	2.64975475
14.62	0.380	0.38167941	3.30307E-06	0.06840366	2.63253515
15.17	0.381	0.384177235	8.98949E-06	0.06591827	2.62343947
15.59	0.383	0.385968201	1.01746E-05	0.06415603	2.61247741
15.90	0.384	0.387266507	1.24776E-05	0.06288873	2.60597085

Tab. 2: Complete results for this calculation.

The next experimental section is about BET-Isotherm and Specific Surface Area where other spreadsheets get calculated.

I.3.2. Specific Surface Area and BET - "Shale 1"

The values of the BET measurement on "Shale 1" are listed in Table 3.
The BET equation

$$\frac{P}{m\cdot(P^*-P)} = \frac{1}{m_{mono}\cdot c} + \frac{c-1}{m_{mono}\cdot c}\cdot\frac{P}{P^*} \tag{18}$$

is transformed to

$$\frac{1}{m\cdot(\frac{P^*}{P}-1)} = \frac{1}{m_{mono}\cdot c} + \frac{c-1}{m_{mono}\cdot c}\cdot\frac{P}{P^*} \tag{19}$$

and the values of $\dfrac{1}{m(\frac{P^*}{P}-1)}$ are plotted vs. $\dfrac{P}{P^*}$ in Figure 3.

Relative Pressure (P/P*)	1/[m(P*/P - 1)]
0.050984646	0.026635464
0.073164220	0.036913625
0.099510463	0.048892608
0.124527145	0.060151338
0.149465958	0.071299507
0.174504900	0.082595229
0.199593911	0.094096754
0.224732982	0.105783369
0.249777491	0.117695800
0.274938813	0.129998953
0.300139079	0.142661694

Tab. 3: Experimental values of BET sorption experiment for "Shale 1".

The BET isotherm for example "Shale 1" yields a straight line in the range of 0.05< P/P*<0.35. This is shown in Figure 7.

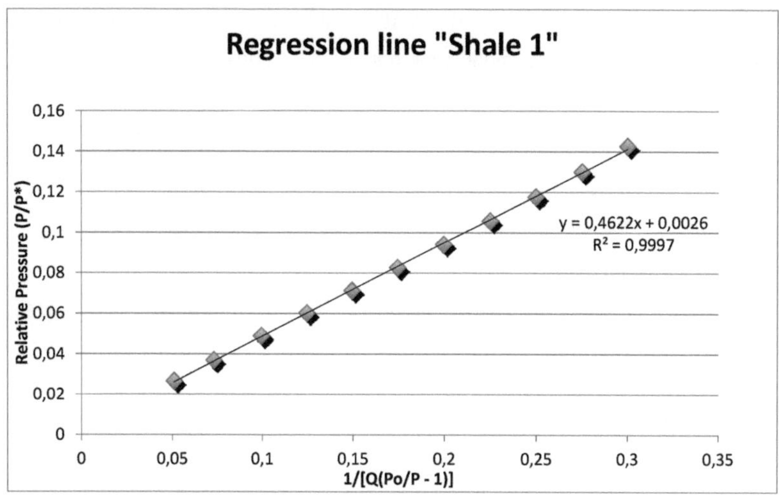

Fig. 7: Regression line for the example "Shale 1". The gradient is y = 0.4622x + 0.0026 with R² = 0.9997. The slope is exactly 0.462190916 and the intercept exactly 0.002608966.

With the slope and the intercept of the regression line one can now determine c and m_{mono} of the BET formula shown above:

$$m_{mono} \cdot c = \frac{1}{0.002608966} = 383.2935662 \tag{20}$$

$$c = 0.462190916 \cdot 383.2935662 + 1 = 178.1548044 \tag{21}$$

$$m_{mono} = \frac{383.2935662}{c} = 2.15146354 \frac{cm^3}{g} \, STP \tag{22}$$

m_{mono} can now convert into the amount of substance (n_{mono}) by using the ideal gas law:

$$p \cdot V = n \cdot R \cdot T \rightarrow n = \frac{p \cdot V}{R \cdot T} \rightarrow \frac{V}{n} = \frac{R \cdot T}{p} = 0.02241407 \frac{m^3}{mol} = 22414.07032 \frac{cm^3}{mol} \tag{23}$$

With R = 8.3145 J/(mol·K), T = 273,15 K and p = 101325 Pa the outcome for V/n is 0.02241407 m³/mol (22414,07032 cm³/mol). Hence the monlayer amount of substance (in mol) adsorbed n_{mono} in per gram of sample is:

$$n_{mono} = \frac{m_{mono}[\frac{cm^3}{g}]}{\frac{V}{v}[\frac{cm^3}{mol}]} = \frac{2.15146354}{22414.07032} = 9.59872 \cdot 10^{-5} [\frac{mol}{g}] \tag{24}$$

With the Avogadro constant (NA) $6.022 \cdot 10^{23}$ mol^{-1} you can determine how many molecules are sorbed in a monolayer per gram of shale (monolayer sorption capacity):

$$n_{mono} \left[\frac{mol}{g}\right] \cdot NA \frac{1}{mol} = 9.59872 \cdot 10^{-5} \cdot 6.022 \cdot 10^{23} = 5.78 \cdot 10^{19} g^{-1} \qquad (25)$$

Taking the molecular cross-sectional area for N_2 (0.1620 nm² or $1.62 \cdot 10^{-19}$ m²) and the n_{mono} value, the specific surface area can be determined. For "Shale 1" we obtain the following value:

$$SSA = 5.78 \cdot 10^{19} g^{-1} \cdot 1.62 \cdot 10^{-19} m^2 = 9.36 \frac{m^2}{g} \qquad (26)$$

I.3.3. Specific Surface Area and BET - "Norit AC"

Another example was the excel-sheet "Norit AC", but here you have the problem, that you just get a straight line by using three values, as shown in figure 8. The given values are shown in table 4.

Relative Pressure (P/P*)	1/[Q(Po/P - 1)]
0.0099	$5.36761 \cdot 10^{-5}$
0.0594	0.000297790
0.1089	0.000556340
0.1584	0.000841698
0.2079	0.001160351
0.2574	0.001519657
0.3068	0.001927728
0.3563	0.002398054
0.4057	0.002944546
0.4552	0.003590398
0.5047	0.004363821
0.5541	0.005305686
0.6036	0.006482765
0.6530	0.007991016
0.7025	0.010001459
0.7519	0.012803417
0.8014	0.017001107
0.8508	0.023953880
0.9003	0.037794190
0.9498	0.078772628

Tab. 4: Given values for "Norit AC".

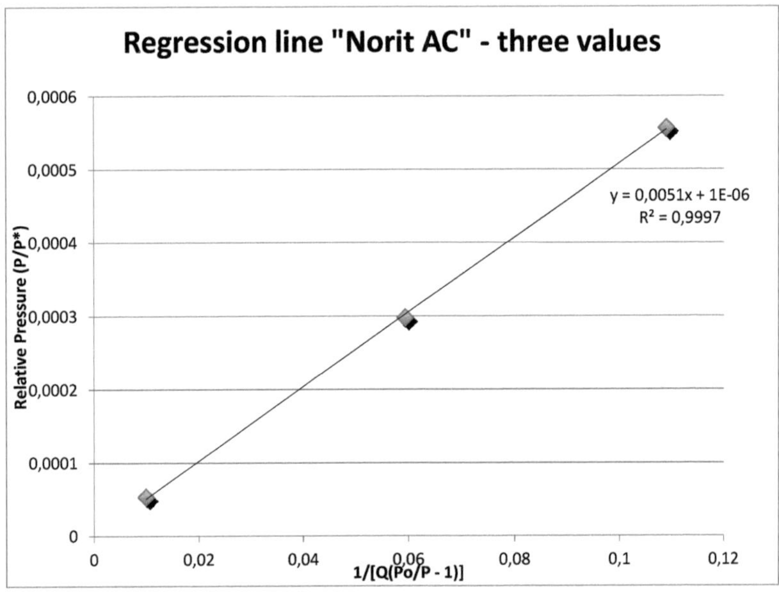

Fig. 8: Regression line for the sheet "Norit AC" by using three of the given values. The gradient is $y = 0.0051x$ (slope) $+ E{-}06$ (intercept) with $R^2 = 0.9997$.

By using all of the given values you won't get a straight line and an accurate regression line is not given as shown in figure 9.

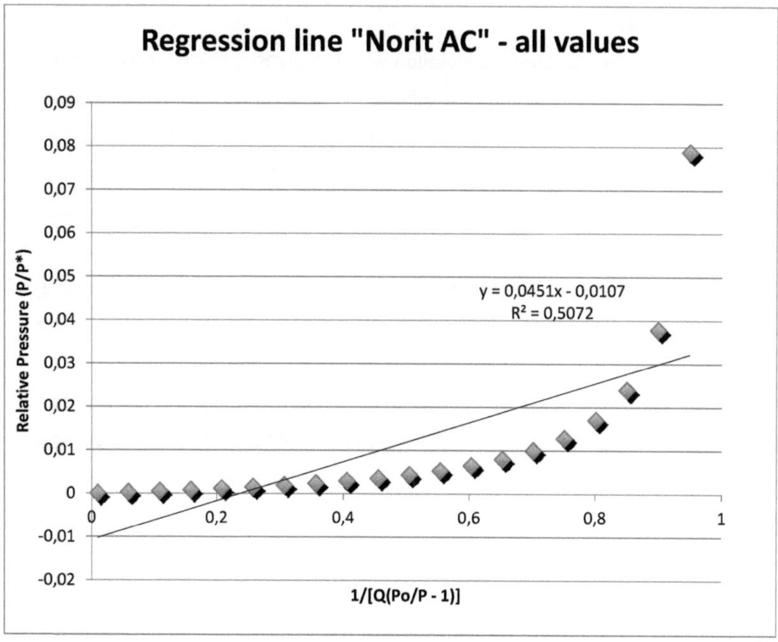

Fig. 9: Regression line for the sheet "Norit AC" by using all of the given values. The gradient is y = 0.0451x (slope) + 0.0107 (intercept) with R^2 = 0.5072.

The three values are not representable for the whole measurement because they are insufficient, so a calculating after the BET method doesn't make sense and the regression line drawn after all values is not accurate enough for a calculation with the BET method.

I.4. Evaluation

The formula for the Langmuir calculation with the given parameter worked good, but P_L and m_{inf} get calculated in a different way. One can do this in two ways. The first one is the solver function in excel, the second one is the calculation of steep and slope with aid of a partial regression line. These values are compared with each other and one see that this solutions are equal. The applied values demonstrate a Langmuir-Isotherm, which looks like the isotherm of Brunauer et al. (1938). The Langmuir-isotherm is the simplest sorption model for monolayer sorption and is a basis for further models, for example the BET-method, which is a type II Langmuir-isotherm and hence for multilayer materials.

Nearly the same way was the working with the BET-method and the calculation of the specific surface area: With a right measurement the calculation was simple and comprehensible (see I.3.2. Specific Surface Area and BET - "Shale 1"), but with wrong measurements or errors in the measurement, a calculation was useless or nearly impossible (see I.3.3. Specific Surface Area and BET - "Norit AC"). Given that the calculation for the specific surface area is built on a right BET-calculation, you have the same problem if the measurement is wrong and therefore the BET-calculation has errors. But in the other case, if you have done a wrong BET-calculation, you have no way to check your values, like you have with the Langmuir-equation.

I.5. References

[1] Sheets of the lecture "Analysis of Microstructural and Petrophysical Data" WS 12/13 by B. Krooss

[2] http://en.wikipedia.org/wiki/Specific_surface_area (05.01.2013)

[3] Brunauer, S. , Emmett, P. H., Teller, E., 1938. Adsorption of Gases in Multimolecular   Layers. J. Am. Chem. Soc., 60 (2), 309–319

[4] http://www.pci.tu-bs.de/aggericke/PC5-Grenzf/Festkoerperoberflaechen.pdf

[5] http://www.chemistrylearning.com/langmuir-adsorption-isotherm/

[6] http://upload.wikimedia.org/wikipedia/commons/8/82/Langmuir_isotherm2.jpg
(14.01.13)

[7] http://de.wikipedia.org/wiki/Sorptionsisotherme#Langmuir-Isotherme
(14.01.13)

[8] http://www.chemgapedia.de/vsengine/vlu/vsc/de/ch/13/vlu/thermodyn/phasen
/phasen_gesamt.vlu/Page/vsc/de/ch/13/pc/thermodyn/phasen/langmuir.
vscml.html (14.01.13)

[9] http://www.pci.tu-bs.de/aggericke/PC5-Grenzf/Festkoerperoberflaechen.pdf
(14.01.13)

[10] http://www.uni-
due.de/imperia/md/content/verfahrenstechnik/dissertation_pahl.pdf
(14.01.13)

[11]http://www.chemgapedia.de/vsengine/vlu/vsc/de/ch/10/adsorption/grundlagen
/grundlagen_der_adsorption.vlu/Page/vsc/de/ch/10/adsorption/grundlagen
/adsorptionsgleichgewichte/langmuir_isotherme.vscml.html (14.01.13)

[12] http://www.gmehling.chemie.uni-oldenburg.de/Praktikum/Adsorption.pdf
(14.01.13)

[13] http://www.chemie.uni-bayreuth.de/mcii/de/lehre/23489/iv_bet.pdf (14.01.13)

[14] http://infohost.nmt.edu/~jaltig/Langmuir.pdf (14.01.13)

[15] http://www.thuisexperimenteren.nl/infopages/langmuir/langmuir.htm
(14.01.13)

[16] http://www.chemistrylearning.com/langmuir-adsorption-isotherm/ (14.01.13)

[18] http://iwan.chem.tu-berlin.de/~lehre/pc/pr1_WS/skript/18_adiso.pdf
(14.01.13)

[19] Itodo, A. U., Itodo, H. U. and Gafar, M. K.; Estimation of Specific Surface Area using Langmuir Isotherm Method; Journal for Applied Science Environmental Manage (December 2010), Vol.14, page 141-145